S0-ANP-671

Collecting for the City Naturalist

LOIS J. HUSSEY and
CATHERINE PESSINO

Collecting for the City Naturalist

Illustrated by BARBARA NEILL

THOMAS Y. CROWELL COMPANY
NEW YORK

BY THE AUTHORS

Collecting Cocoons
Collecting Small Fossils
Collecting for the City
Naturalist

Published simultaneously in Canada by Fitzhenry & Whiteside Limited, Toronto.

Designed by Mina Baylis Greenstein

Manufactured in the United States of America

ISBN 0-690-00317-X

Library of Congress Cataloging in Publication Data
Hussey, Lois Jackson, date
Collecting for the city naturalist.
SUMMARY: Suggestions for exploring such city environments as streets, vacant lots, and ponds and for making collections from the natural specimens found there.
1. Natural history—Technique—Juv. lit. [1. Nature study] I. Pessino, Catherine, joint author. II. Neill, Barbara, illus. III. Title.
QH48.H83 500.9'173'2 73-17293
ISBN 0-690-00317-X

1 2 3 4 5 6 7 8 9 10

Contents

Discovering Nature in the City

A city, to most people, is a place of streets with tall buildings, lots of people, and traffic. It is an unnatural place made of brick, concrete, and steel. For most people the only animals are dogs and pigeons; the only plants are those growing in flowerpots and window boxes.

But there is much to discover about animals and plants and rocks in a city if you know how and where to look. The very materials used in building a city come from the earth. There are small worlds of plants and animals living in the cracks of the sidewalks, in the trees along the streets, in yards, vacant lots, playgrounds, and parks.

You can explore these different places. They can be your outdoor laboratories. You can make discoveries the same way naturalists do. Naturalists study the natural world of plants, animals, and rocks. They observe, describe, compare, measure, experiment, ask questions, read, keep records, photograph, and collect specimens. You can too.

In this book are suggestions for exploring the different environments of a city and directions for making collections of what you find there.

Many naturalists keep notebooks in which they record the things they see each day. These notebooks, filled with their daily observations, are called journals. Many started keeping journals while very young and have used the carefully recorded information years later in their scientific work and in their books. One famous naturalist, Edwin Way Teale, tells in his book *Dune Boy* how he started to collect information before he was eight years old, recording observations in pocket notebooks about animals, plants, rocks, and the weather.

Scientists use the word *data,* meaning facts, for such information. An important part of a scientist's work is collecting data learned by direct observation. Carefully recorded data may also be of help to other scientists.

Recording Data — Where To Begin

Start by observing one thing each day. It can be something in your yard, the lot nearby, on the street, in the park, or anywhere you see something that in-

terests you. Write what you observe in a notebook. Use a bound notebook as loose sheets of paper may be lost.

What To Record

Write down all that you can about what you are observing. Many facts about common city plants and animals are not known. The "unimportant" may turn out to be important.

Do not wait too long to write down what you see. You may forget some details.

Write or print clearly. You will want to read your notes later on.

With each observation record the date, place, time, and weather.

Place

Draw a map of your neighborhood to show the location of your home, school, playground, park, and the other areas where you make your observations. The map will help locate the exact place an observation was made.

Scientists often make a grid of an area under study. Using surveyors' instruments, they divide the area into squares called quadrats. Posts are placed at the corners of the quadrats. Each post has its own letter and number. In the example shown here letters are used to mark off the lines running east and west and numbers for the lines running north and south. By

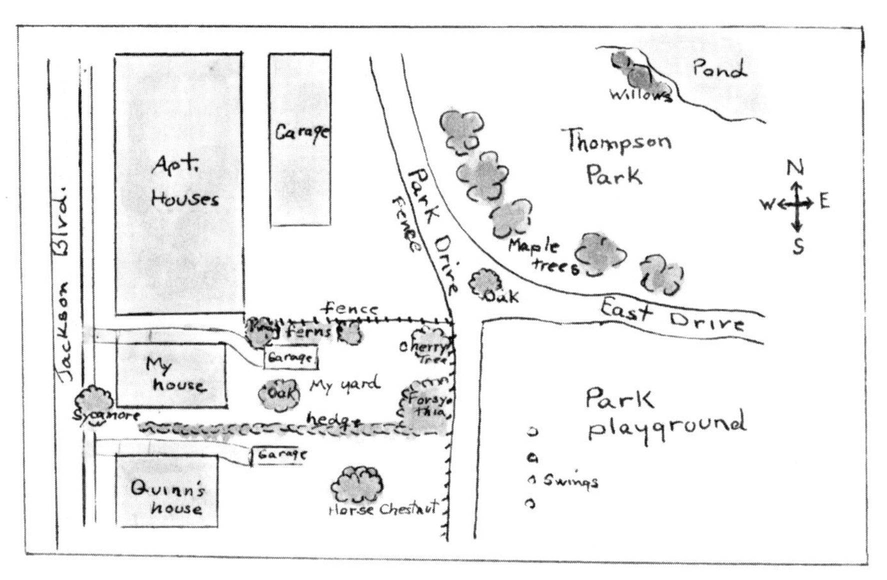

dividing an area into smaller quadrats, the scientist can more easily record exact locations.

Scientists use the metric system in taking measurements. One meter equals 39.37 inches. If you decide to make a grid of an area, you might make each quadrat one meter square – 39.37 inches long on each of the four sides. The United States is gradually changing over to the metric system.

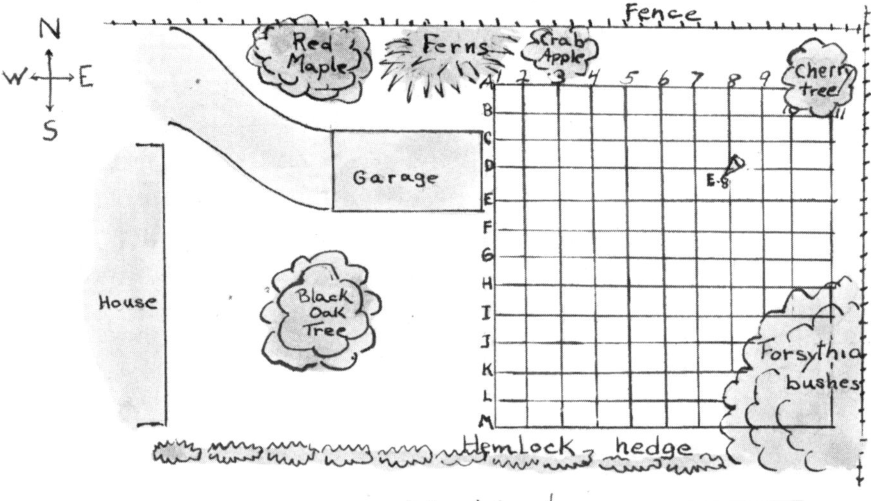

My Yard

Time

To avoid errors the 24-hour clock is used in recording data. It may seem strange at first, but can you see why it is a better clock to use? Scientists, military personnel, airlines, railroads, and others use it.

The following examples will help you translate a 12-hour clock into a 24-hour one.

1:00 a.m. —0100	*1:00 p.m.—1300*
2:00 a.m. —0200	*2:00 p.m.—1400*
3:15 a.m. —0315	*3:15 p.m.—1515*
6:30 a.m. —0630	*6:30 p.m.—1830*
10:05 a.m. —1005	*10:05 p.m.—2205*
12:00 noon —1200	*midnight—2400*
12:30 p.m. —1230	*12:30 a.m.—0030*

To express the time in words you say 0-one-hundred hours, 0-three-fifteen hours, ten-0-five hours, twelve-thirty hours, fifteen-fifteen hours, twenty-four-hundred hours, 0-0-thirty hours.

Weather

Is it clear, sunny, cloudy, foggy? Is it raining or snowing? What is the temperature? Is it calm or is there a wind? What is the wind direction? How hard is the wind blowing?

As you add to your records, you will be able to see patterns in plant and animal activities. These patterns are related to the time of year, the time of day, and the weather. This is why it is important to record weather conditions. Weather also affects rocks, buildings, and pavements.

Larger cities have their own weather stations that report over the radio and TV and in the newspapers. In these reports the weather is often given as that taken at the local airport or weather station.

But weather varies locally. For example, the temperature and wind in your neighborhood may be very different from that reported by the nearest weather station.

Beaufort Scale	*Observation*	*Description*	*Wind Velocity*
0	Smoke rises straight up. Nothing moves.	calm	under 1mph
1	Smoke drifts slowly. Tree leaves barely move.	light airs	1-3mph
2	Leaves rustle. Wind felt on face.	slight breeze	4-7mph
3	Leaves and twigs move. Loose paper and dust raised from the ground.	gentle breeze	8-12mph
4	Small branches move. Dust and paper raised and driven along.	moderate breeze	13-18mph
5	Small trees begin to sway. Small waves on pond.	fresh breeze	19-24mph

An outdoor thermometer will give the exact temperature.

You can tell the wind direction if you know which way is North. Remember, the wind direction is the direction from which the wind is blowing.

The *Beaufort scale*, named after the British Admiral, Sir Francis Beaufort, who invented it, indicates the speed or *velocity* of the wind. The velocity is given in miles per hour (mph), the number of miles the wind travels in one hour.

Beaufort Scale	*Observation*	*Description*	*Wind Velocity*
6	Large branches sway. Umbrellas used with difficulty.	strong breeze	25-31mph
7	Whole trees in motion. Difficult walking against the wind.	moderate gale	32-38mph
8	Tree twigs break. Walking progress slow.	fresh gale	39-46mph
9	Branches break. Slight damage to structures.	strong gale	47-54mph
10	Trees snap and are blown down. Considerable damage to structures.	whole gale	55-63mph
11	Widespread damage.	storm	64-72mph
12	Extreme damage and destruction.	hurricane	above 72mph

Standard symbols from weather maps are easy to use in recording weather.

○ *clear*	◑ *partly cloudy*	● *cloudy*	Ⓕ *fog*
Ⓡ *rain*	Ⓩ *freezing rain*	Ⓣ *thunderstorms*	Ⓢ *snow*

Ground conditions are an important part of the total weather picture. Is there ice on puddles or pond? Snow cover on the ground? Ice and snow melting? Flooding?

Sketching

Sketches are another way of recording observations. Making sketches will help sharpen your powers of observation.

Use separate sheets of paper or a sketch pad if you don't want to sketch directly in your notebook. Work quickly and freely, without erasing. You will improve with practice. Beside the sketches jot down notes on color, size, and other details.

Sketches can be of just as much help in recording observations about plants and rocks as about animals.

Photography

Photographs can be excellent records also. They make it possible to see details you might have missed. They make it easy to share your observations with

others. Even with a simple camera you can obtain good results.

Here are some suggestions for subjects you might wish to record:

pigeon strutting
friend feeding ducks at a pond
gulls at a garbage dump
sparrows taking a dust bath
tree flowers
weeds in pavement crack
cloud formation
rock specimens
squirrel on a branch

Sound Recordings

Another tool of the scientist is the tape recorder. Scientists use very elaborate recording equipment, but you will be able to record the sounds of nature with any type of recorder you have or can borrow.

At first you may hear only the noise of traffic and

people. But listen carefully. You will be able to hear the sound of rain, wind in the trees, a dog barking, a pigeon cooing, a gull overhead, a jay calling, a fly buzzing, a cricket chirping, a squirrel scolding, and many others.

You can learn the sounds made by animals by listening to recordings. Knowing their sounds is a help in identifying the animals in the field.

What happens when you play back sounds made by an animal? How does the animal react? Do other animals come to the sound?

Along the Street

COLLECTING ROCKS AND MINERALS

The rocks used in building the city can be the beginning of a rock collection. All come from the earth. Some, such as granite, marble, slate, sandstone, and limestone, are used in their natural state. Others are refashioned.

Clay, a part of many building materials, is decomposed (broken-down) rock. Brick and tile are made from clay. Cement is made from clay and limestone. Cinder blocks are made from cement and cinders; concrete from cement, sand, and gravel; terrazzo from marble chips and cement. Plaster is made from gypsum or lime and sand; glass from sand, lime, and soda. Steel is made from iron.

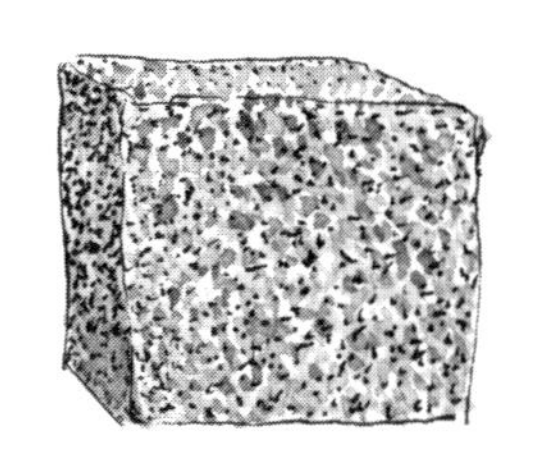

Often the building stones used in a city are made nearby or are mined in nearby quarries. It may be possible for you to visit a quarry, a stonemason's yard, or a monument works to see men working with different kinds of stone.

For an important building, stones may be carried great distances. Vermont granite, Italian marble, Texas limestone, are building stones used in cities all over the country because of their beauty and

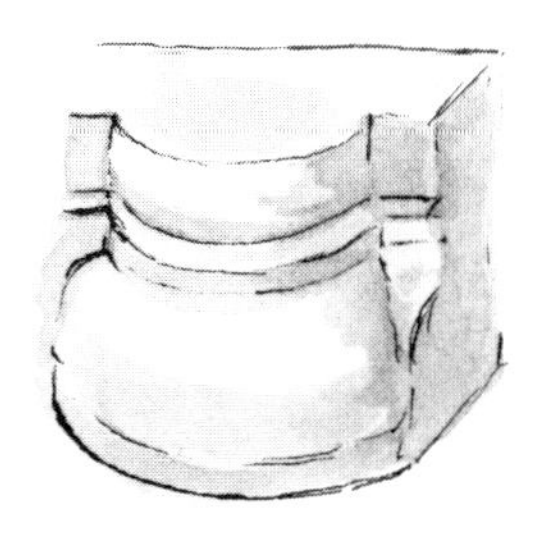

strength. Much can be learned about rocks just by looking at buildings.

You cannot collect specimens from buildings that are standing, but you can collect from buildings that are being torn down. This is always going on in a city. Much can be found in the rubble left behind.

Rocks are composed of minerals. Four minerals commonly found in rocks are quartz, feldspar, mica, and hornblende. A rock such as granite contains as many as ten different minerals. Sandstone is made up almost entirely of one mineral.

The best rock and mineral specimens collected in cities usually come from excavations for buildings, subways, utilities, and highways. The excavations uncover the rock.

The rocks under a city are often visible as rock outcroppings. Here is an illustration of one along a street. The building wall rests on the outcropping.

Rock outcroppings may be seen and studied more easily in city parks. In most parks collecting is forbidden. Rocks, as well as plants and animals, should be left for all to enjoy.

But often loose pieces of rock fall off as a result of wear and weathering. These loose pieces are called "float" by geologists (scientists who study rocks). No one seems to object to float being collected.

You can buy rock and mineral specimens, but the greatest pleasure is in finding your own. Once you have a good-sized collection, you can swap specimens with other collectors.

Remember, if you wish to explore on other people's property, get the owner's permission first. Be sure to follow any rules the owner makes.

EQUIPMENT

Notebook and pencil to record data.

Tape to mark specimens in the field.

Newspaper to wrap specimens.

Geologist's pick and cold chisel to break large pieces of rock into smaller specimens; specimens 2 by 3 inches are easy to handle.

Safety goggles to protect your eyes when breaking rock.

Heavy gloves to protect your hands.

Hand lens (10 power) for a closer look.

Knapsack.

The pick, cold chisel, and safety goggles may be bought at a hardware store, and a hand lens at an optician's shop; or they may all be ordered from a scientific supply house. See the list on page 67.

Collecting Rocks

If you can collect from an outcropping, you will need to use the pick and cold chisel. Remember always to aim the point of the chisel away from you when breaking rock.

Surfaces of rocks are changed by rain and wind or may be covered with soot. It is necessary to break the rock with a few sharp blows of the pick to see a fresh surface.

Recording

Stick a small piece of tape on the specimen. Write the number of the specimen on the tape. If it is your first specimen, it will be number 1. In your notebook write number 1. Next to it record the following: date, name of specimen (if you know it), the exact place

where you found it, whether it was float, rubble, or taken from a rock outcropping.

After recording the data in your notebook, wrap the specimen in newspaper, fasten with tape, and place in your knapsack.

At home use warm, soapy water and an old toothbrush to clean your specimen. Dry thoroughly.

Keeping a Collection

Each specimen in a collection should have a permanent number. Paint a small circle with white or yellow enamel on the specimen. After the paint dries, write the specimen number on the circle with permanent ink. After the ink dries, coat the circle with clear nail polish or clear shellac to prevent the number from wearing off.

A geologist transfers the data from his field notebook to permanent record cards. You may want to do the same. Three-by-five-inch index cards are suitable. Use one card for each specimen.

Egg cartons, cigar boxes, shoe boxes, greeting card boxes, plastic boxes, and other containers can all be used for keeping and displaying a collection.

Grouping

At first you may want to keep the specimens in your collection in the order in which you found them or according to where you found them. But as you learn more about rocks, you will probably want to arrange them according to the way they were formed or by their mineral content.

In the bibliography you will find a list of books useful in identifying rocks and minerals.

Collecting Sand

Some "rockhounds" we know collect sand. There are many kinds of sand. Sand from a beach is different from sand found in a lot, a riverbed, or weathered building stone. Sand from one beach may be different from sand found on another beach.

Sand is broken-down rock. It is made up largely of quartz, but crystals and pieces of other minerals also show up. This is what makes sand collecting fun. You can have a miniature mineral collection in only one handful of sand.

To really see sand you need to look at it through a magnifying glass or a hand lens. Then you will be able to see the different shapes and colors of the individual grains. The different minerals give each sand sample its color.

Small jars will hold your collection.

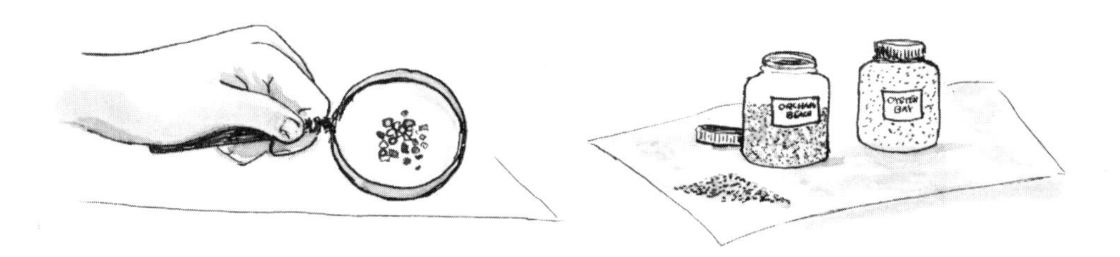

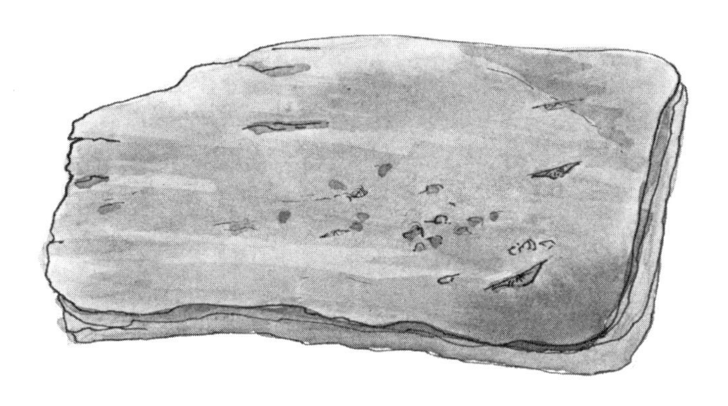

Collecting Fossils

In some cities fossils are commonly found in the rocks. The fossils are the remains or other evidences of animals and plants that lived a long time ago.

If you live where fossils are common, you may want to collect them. Use the same equipment and techniques used in collecting rocks and minerals.

If you live where fossils are not common, you may be able to see them in building stones such as shale, sandstone, and limestone. These stones sometimes contain fossils.

Collecting Snowflakes

Water is classified as a mineral. Water occurs as a liquid, or a solid (ice), or as a gas (water vapor). When conditions are right, water vapor crystallizes and falls as snow.

Sometimes snow is fine and dry. Sometimes it is wet and heavy. Sometimes snow falls as big, fluffy flakes. Have you ever examined a snowflake? No two flakes are exactly alike.

To collect snowflakes you will need glass slides or any small pieces of glass, a flat piece of wood to hold the glass, an eye dropper, and a 2 percent solution of polyvinyl formal resin dissolved in ethylene dichloride. Ask at a drugstore for this.

1. Place solution and glass out-of-doors in a spot protected from dirt and falling snow. Both the glass and the solution must be very cold before you can collect snowflakes.
2. When cold, using the eye dropper, cover the glass with the solution.
3. Carefully catch a few snowflakes on the coated glass. Put the glass in a spot protected from falling snow.
4. If any of the snowflakes are not completely covered by the solution, place a small drop of the solution on the flake.
5. Let dry for 15 minutes.

When dry, the glass can be taken indoors. Examine the impressions of the snowflakes with a magnifying glass or a hand lens. Handle gently. The impressions are delicate. Store your collection in a small box.

The Tree on the Street

COLLECTING PLANTS

A tree growing on your street or in front of your school can be your introduction to botany. Botany is the study of plant life. The scientist who studies botany is called a botanist.

Begin by making observations. What is the shape of the leaf? Does the tree drop its leaves at the end of each growing season? When do new leaves come out? Does the tree have flowers? When do the flowers bloom? Unless the flowers are showy such as those of a magnolia tree, they may not be noticed. What does the fruit look like? The seed container, whether it is fleshy or dry, is called a fruit by botanists. What do the seeds look like? These are some of the observations that can be made and recorded in your field notebook.

It would be interesting to pick one tree and keep a record of it throughout a year.

Botanists preserve plants by pressing and drying them. They use plant presses, special mounting papers, and other materials bought from scientific supply houses. These materials are not necessary. You have at home materials you can use.

Here are some suggestions for plant collections.

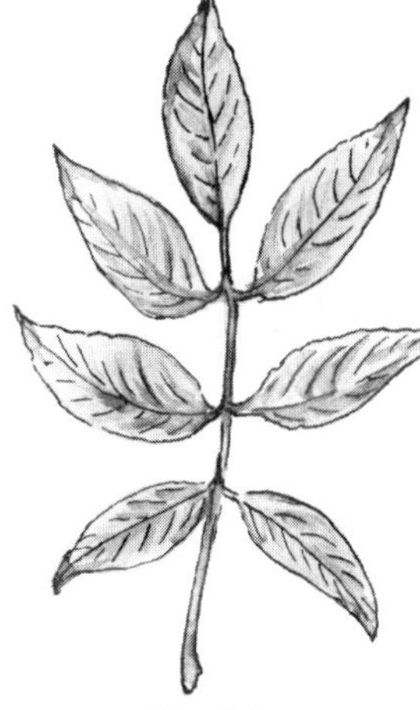
White Ash

Norway Maple

Chokecherry

Eucalyptus

How to Mount a Leaf

Collect a leaf when it is fully grown. Collect one that has not been eaten by insects. Choose one that will fit on your mounting paper.

Carefully remove the leaf from the tree, petiole (stem) and all. Place the leaf in a plastic bag and seal to prevent drying if you are far from home. At home:

1. Carefully lay leaf specimen between a folded sheet of newspaper.
2. Add 15 or more layers of newspaper under the folded sheet and 15 more on top of it.
3. Cover with a board or heavy cardboard the same size as the newspaper.
4. Weigh down with books or rocks.
5. Change the layers of newspaper under and on top of the folded sheet every day for one week, then every other day for ten days. The leaf should then be dry.

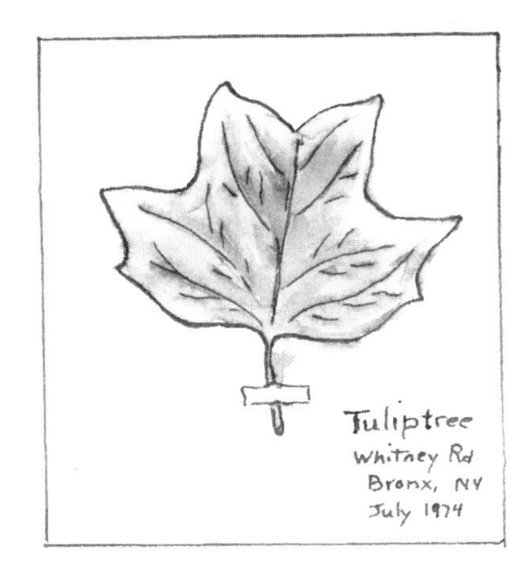

6. Glue dried specimen to mounting paper. Large index cards, cardboard, construction paper, and scrapbooks can be used for mounting leaves. Any glue can be used. Thick petioles can be held with strips of tape.

7. Label with name and location of tree and date.

How to Make Ink Pad Prints of Leaves

If several of your friends or your club or class become interested in making a leaf collection, it may be best to make ink pad prints. One leaf can be used by everyone. In this way a tree will not be stripped of its leaves, which are needed by the tree to make its food.

For this you will need a leaf, ink pad, newspaper cut to the size of the pad, tweezers, and plain paper.

1. Place a leaf, underside down, on the ink pad.
2. Place a piece of newspaper over the leaf.

3. Carefully rub your fingers over the newspaper, pressing all of the leaf against the ink pad, to coat the leaf with ink evenly.

4. Remove newspaper.

5. Pick up the leaf with tweezers and place ink-side down on plain paper.

6. Cover with a clean piece of newspaper.

7. Hold newspaper and leaf firmly with one hand. With the other hand, rub your fingers over the newspaper, pressing all of the leaf against the paper. Do not move the leaf while doing this or the print will be smudged.

8. Remove newspaper.

9. Use tweezers to lift leaf from the paper.

10. Let the print dry. Label.

How to Make a Leaf Skeleton

The delicate veins of a leaf can be seen by removing the fleshy part of the leaf. It is through the veins that raw materials (minerals and water) needed to make food reach all parts of the leaf. The food manufactured by the leaf is then carried through the veins back to the rest of the tree.

Making a leaf skeleton is hard, but it should be done at least once to see the network of veins. Use a fresh green leaf with tough veins, such as an oak leaf. You will need an old piece of carpet or several thicknesses of felt or soft cloth and a hairbrush or shoebrush. A brush with natural bristles all the same length is best.

1. Place leaf on cloth or carpet.
2. Hold leaf firmly with one hand.
3. Tap leaf with brush held in other hand.
4. Every once in a while turn leaf over and tap other side.
5. Keep tapping until only the veins are left.

To preserve the leaf skeleton, dry between layers of newspaper weighted down with a book. Mount the dried leaf skeleton on paper or between two pieces of glass held together with tape. Label.

Collecting Tree Twigs

If you live where you have winter or a dry season when trees drop their leaves, you will be able to make a twig collection. Here are the parts of a twig.

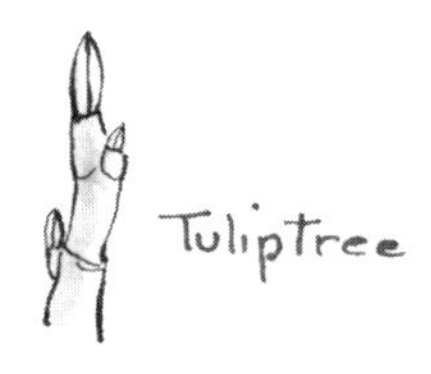

Buds contain tiny leaves or flowers or both. *Bud scales* protect the leaves and flowers from drying. Notice the size, shape, and color of the buds. Are the buds smooth, sticky, hairy?

Leaf scars remain when leaves drop off. Are the leaf scars opposite each other or are they alternate?

Bundle scars (the dots on leaf scars) are the scars of veins through which sap (minerals and water) flowed from the roots to the leaves, and food made

by the leaves was carried back to the rest of the tree. How many bundle scars are there on each leaf scar? How are they arranged?

Lenticels are places in the bark through which the tree "breathes."

The *pith* is the soft center of the twig.

Trees can be identified by their twigs.

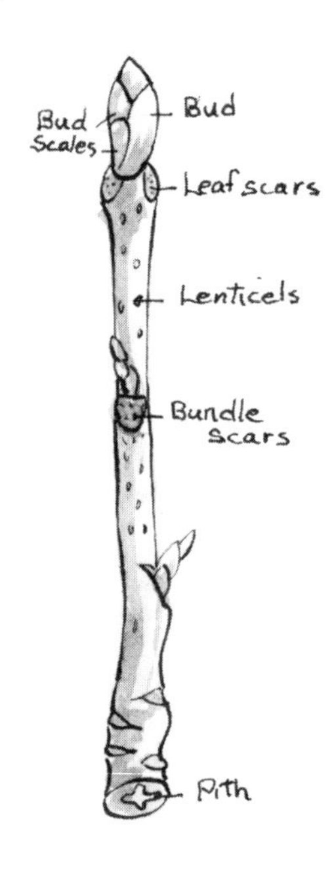

How to Keep a Twig Collection

Using shears, cut a twig from a tree. Tape the twig to a large index card or tie it on using a needle and thread. Label. On the card note the color of the twig and buds, and whether they are hairy or sticky. These characteristics may change or be lost as the twig dries. If it is a delicate twig, protect it by placing the mounted twig in a plastic bag.

A twig collection can be stored in a shoe box or small carton.

Tree Shapes

Many kinds of trees are easily recognized by their shape. In a city, conditions may alter the shape.

Notice how trees along a street lean away from the buildings to get more light.

Low branches that interfere with traffic are removed. Top branches are removed to make room for overhead wires.

People and vehicles damage trees, sometimes causing them to grow crookedly.

Ginkgo

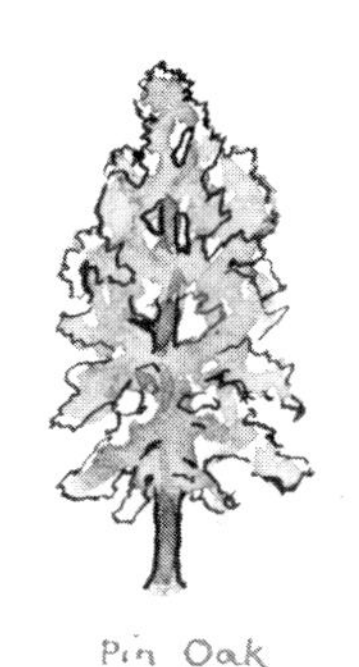

Pin Oak

London Plane

Washington Palm

Lack of space for roots, and lack of soil and water, can cause stunted growth.

Polluted air also can cause stunted growth.

Run-off from rock salt spread on sidewalks "poisons" the soil, causing stunted growth.

Dog urine "burns" the bark of young trees, causing scars to form.

In open areas, such as parks, trees are more apt to grow fuller.

Tree shapes can be collected by sketching or photographing them.

How to Make a Bark Rubbing

Trees can be identified by their bark. Removing bark spoils the looks of a tree, injures, and may kill the tree. A bark rubbing can be made without harming a tree.

To make a bark rubbing, tape a piece of plain paper to a tree trunk. Rub a fat crayon held on its side against the paper. Once you have completed the rubbing, remove the paper. Label.

If you examine the tree carefully, you may find "stories" to record with rubbings.

The city Parks Department is usually responsible for taking care of the trees it plants. During the year, the Parks Department employees go through neighborhoods checking trees. Sick or damaged branches, branches that are too low, or ones that interfere with overhead wires are removed.

You might tell a Parks Department employee about your interest in botany and ask for samples (six inches of a fat branch) for a bark collection. The employees we have asked have always been willing to help.

Use sandpaper to smooth one of the ends. This will make it easier to see the annual growth rings. Each annual ring consists of an inner layer and an outer layer. The inner layer is made of wood formed in the spring, the outer layer is of wood formed in the summer. Counting the number of annual rings will tell you the age of the branch. The branch will, of course, be younger than the tree.

How to Collect Tree Flowers and Fruits

For the botanist, it is important to collect the flowers and fruits as well as the leaves, twigs, and bark, as all of the parts are used in classifying plants.

Most tree flowers are small and delicate and hard to press. Large, showy flowers are fleshy and hard to dry. It is easiest to collect tree flowers by sketching or photographing them.

The fruits of many trees are hard. They can be picked up, placed in open boxes, and left to dry.

Some fruits will open as they dry and drop their seeds. Tiny seeds can be put in a small envelope and stored with the fruit. Some fruits must be opened to see the seeds inside.

Some fruits you collect may be empty. Their seeds will already have been dropped.

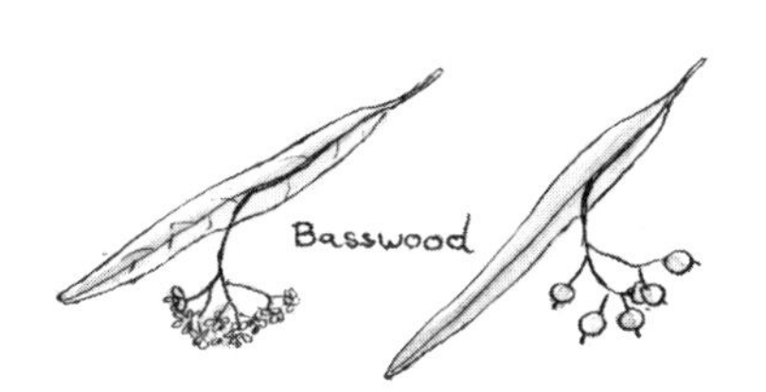

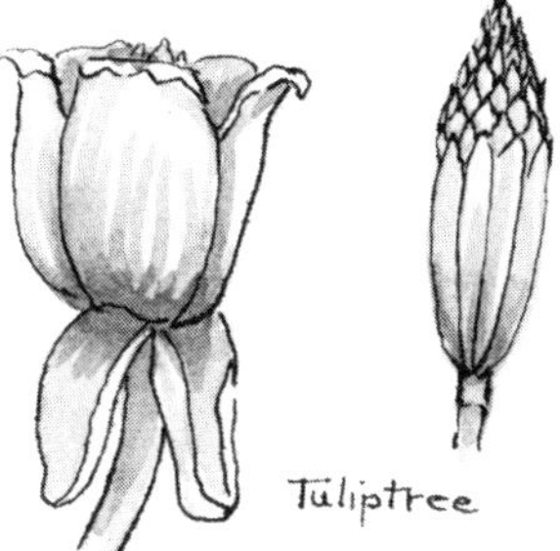

All the Parts—How to Make a Riker Mount

In time you may want to arrange your collections so that all parts of one species (kind) of tree are together. A "Riker mount" that you can make out of a cardboard box, cotton, clear plastic, tape, and straight pins is suitable.

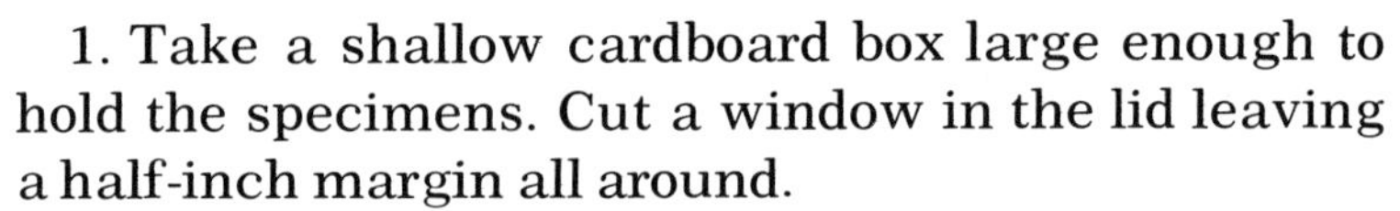

1. Take a shallow cardboard box large enough to hold the specimens. Cut a window in the lid leaving a half-inch margin all around.

2. Place a layer of cotton in the bottom of the box.

3. Arrange the specimens on the cotton. Add or remove cotton so that the specimens are level with the top of the box.

4. Add a label.

5. Cover the window in the lid with clear plastic. Fasten the plastic with tape. Put the lid on the box.

6. Insert straight pins through all four sides to hold the lid in place.

Greeting-card boxes with see-through lids can be used as they are. Most tree specimens though will require larger boxes.

Collecting a Whole Plant—Mounting Botanical Specimens

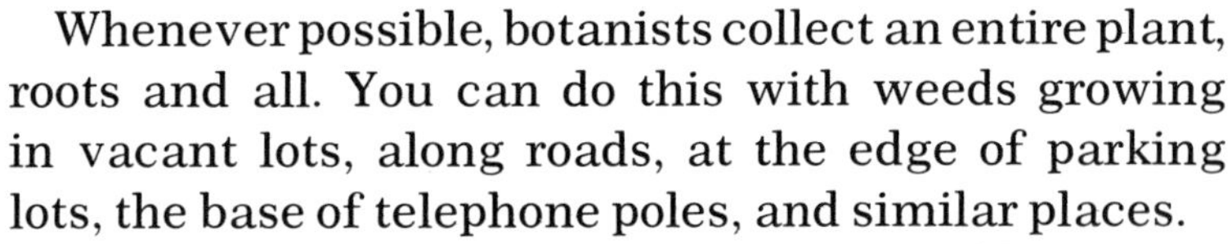

Whenever possible, botanists collect an entire plant, roots and all. You can do this with weeds growing in vacant lots, along roads, at the edge of parking lots, the base of telephone poles, and similar places.

Weeds are an important part of the wild community. They provide food and shelter for wildlife. They produce large numbers of seeds. Weeds are hardy; they grow in waste areas where other kinds of plants cannot survive. They help hold soil and moisture, and some are among our loveliest wildflowers.

Collect the plant when it is in flower. Pull up the

entire plant, roots and all. Shake off loose soil. Place plant in a plastic bag to prevent drying.

At home, wash off any remaining soil around the roots. Blot dry with paper toweling or newspaper. Then follow the instructions for mounting a leaf, turning one leaf so that the underside shows. If the plant is too large for the mounting paper, bend the stem. A very large plant can be cut into two or more sections before drying.

Living Collections

Botanists also keep collections of living plants. These collections are on display in botanical gardens and in arboretums.

You might start such a collection by planting seeds you collect.

All seeds contain the beginnings of new plants, but not all seeds will sprout. Some seeds do not remain "good" for long. Some have special requirements. For example, apple seeds must go through a freezing period before they will sprout. But it is fun to experiment.

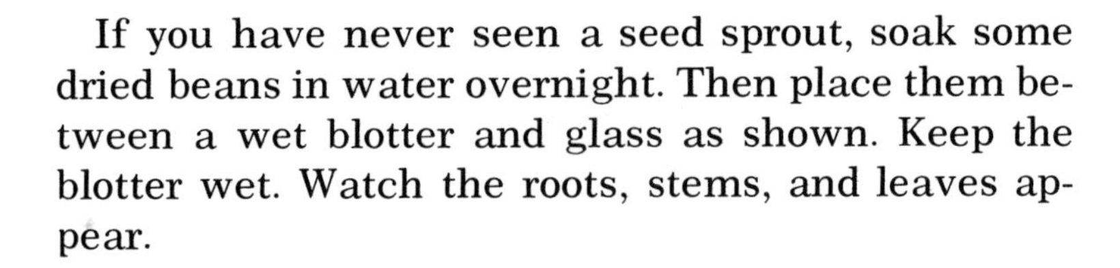

If you have never seen a seed sprout, soak some dried beans in water overnight. Then place them between a wet blotter and glass as shown. Keep the blotter wet. Watch the roots, stems, and leaves appear.

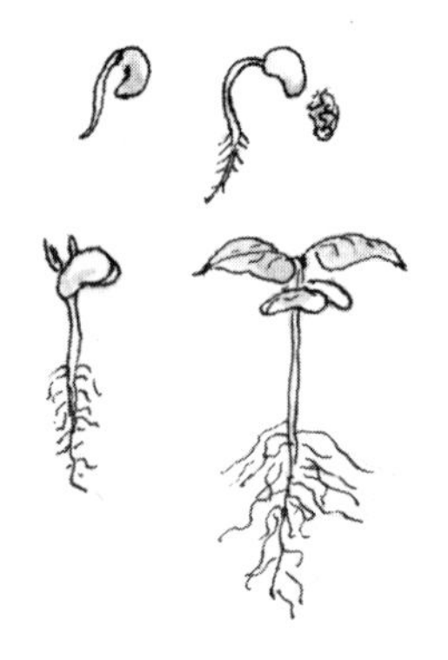

The Kitchen Garden

Look around your kitchen for seeds to plant—dried peas, melon, tomato, grapefruit, orange, lemon. Soak the seeds in water overnight, then plant in small pots nearly filled with soil. In a city it is easiest to buy small amounts of topsoil as needed. The seeds should be covered with about one-half inch of soil. Keep the soil moist. Put the pot in a sunny place.

To start an avocado, place the seed half in water, flat end down, until the roots and shoot appear. This may take as long as two months. When the shoot is two inches long, the seed may be planted.

Some plants grow from stems. A white potato is a stem. Plant a piece of potato with two or more "eyes" (buds).

Some plants grow from roots. A sweet potato is a root. Place one in a jar of water as shown and watch it grow.

In cities where people from all over the world live, it is possible to find fruits and vegetables from other countries for sale. For example, yautia, a vegetable used by people of the Caribbean, produces a lovely houseplant.

The Terrarium

In a city terrarium you can plant the seedlings you raise or small weed plants you collect. You can also buy plants at the dime store or flower shop. Buy small plants that will not outgrow the terrarium in a short time. Choose plants that require the same kind of soil and the same amount of moisture and light. You may need to experiment at first to learn which plants grow best in your terrarium.

To set up a terrarium:

1. Put a layer of pebbles or bits of broken flower pots in first for drainage.
2. Add two or more inches of topsoil.
3. A small hill and a rock in the background add interest.
4. Add plants – smaller in front, larger in back. Be sure roots are covered with soil, and the plant is firmly anchored in place. Do not allow any leaves to touch the glass or they may remain wet and rot.
5. Water thoroughly so that all the soil is moist. Wipe glass clean.
6. Place lid on terrarium. The lid should fit loosely to allow air to circulate.

Remember to keep the terrarium out of direct sunlight and away from radiators. Water lightly when necessary. If water collects on the side of the terrarium, it will be too moist. Remove the lid for a few hours each day until the excess water disappears.

Planting a Street Tree

All cities need more trees. Trees add beauty and interest to a street. Besides casting shade, their leaves cool the air by giving off moisture. Trees help purify the air by removing carbon and replenish the air by giving off oxygen. Trees absorb noise.

In many cities people may buy and plant trees on their street after first getting a permit from the Parks Department. The permit issued in New York City gives information about planting trees. It tells where a hole may be dug so that the tree roots will not interfere with utility lines; what size the hole must be; how much topsoil and fertilizer is needed; and what kinds of trees grow best in the city.

If you would like to plant a tree in your neighborhood, ask your Parks Department for help.

City Birds

COLLECTING DATA

Pigeons come to mind first when we think of city birds. Pigeons were brought here by early settlers to be raised for food. As cities grew, escaping pigeons found shelter and nesting places on window sills and building ledges, which were like the rocky cliffs where they had nested when wild. As their numbers increased, pigeons spread throughout the United States. Today, they are common in the country as well as in cities.

House sparrows and starlings are two other common city birds imported from Europe.

A few house sparrows were released in Brooklyn, New York, in 1853. These birds thrived and multiplied. Others were later released in New England and eastern Canada. By 1875 house sparrows had spread to the Pacific Coast.

A hundred starlings were released in Central Park, New York City, in 1890–1891. They, too, readily adapted, multiplied, and soon spread to the west coast. In a bird count taken recently, in which only small sections of the United States and Canada were covered, thirteen million starlings were counted. There probably are many times that number in North America.

Other species of birds imported from Europe and released were not able to adapt and did not survive.

Bird Migration

Birds, like the pigeon, house sparrow, and starling, that remain in the same area all year are called *residents*. A bird that comes into an area to nest is a *breeder*. A bird that appears only in winter is a *winter visitant*. Birds that periodically move from one area to another (migrate) are *migrants*.

It is the migrants that make ornithology (the study of birds) especially exciting for the city dweller. Many cities are located on one of the four major migration routes. As the amount of green space is limited in a city, migrating birds gather in city parks to feed and rest. It is possible to see a great number of different species as they pause in the parks before moving on.

Central Park, in the middle of New York City, is famous for its spring migration. Birds flying north to their breeding grounds, after a winter spent farther south, pause in the park to rest and feed. On a "wave day," when great numbers of migrants arrive, it is possible to see seventy-five or more species in one day.

Bird Identification

Birds are identified by their size, shape, color, markings, flight, behavior, song, and habitat (where they are usually found). There are several excellent field guides to help you identify birds. Begin by learning the birds outside your window, on your street, and in your neighborhood. City dwellers may have only one tree in a tiny yard, but they can build up a long list of birds seen there.

The only piece of equipment the bird student needs is a pair of field glasses. Opera glasses, which you may be able to borrow, generally magnify two or three times, and do help. But you will soon want stronger field glasses. Good, inexpensive field glasses are available in many stores.

Collecting Bird Lists

As you get to know more birds, you will want to keep track of the ones you identify. You can do this by making lists.

Daily List When you take a bird walk, write down

the names of the birds you see. Some bird students use printed field cards that list the names of birds seen locally. The cards may be bought from bird clubs or from museums. There is space for checking off the species and the number seen and space to record the date, place, time, and weather.

Year List From your daily lists you can make a list of all the birds you see during the year. In New York City nearly 400 species of birds have been seen. A score of 200 in one year is considered very good.

Life List A bird seen for the first time in your life is a "life bird." A life list is a list of these birds. In North America (north of Mexico) it is possible to get a life list of well over 600 species. Stuart Keith, an ornithologist who has traveled widely around the world, has a life list of 4450 species, and he is still adding to his list.

A *Bird Calendar* can be used to record the date you first see a bird during the year. Space can be left on the calendar for recording other information as well.

In time, carefully kept lists may be of scientific importance by giving information on numbers of birds and changes in migration. In *Birds of New York State,* the author, John Bull, used records of "birders" from all over the state, records giving the dates birds were first and last seen in an area, the numbers of birds seen, nesting dates, and other information gained by careful observation.

Field Work

In the beginning you will be identifying the birds you see and building up your lists. In time, you will want to know more about birds.

You may recognize pigeons, but what do you really know about them? Can you tell a male pigeon from a female pigeon? What is the courting behavior? Do they both build the nest? What is the nest made of? How many eggs does a pigeon lay? When do the young pigeons leave the nest? What sounds do pigeons make?

You can learn the answers to most of these questions by observation; or you can learn the answers by reading, as scientists have studied pigeons. Many common birds have not been studied. This is where you can make a real contribution to ornithology. It is a matter of watching birds and carefully recording your observations. Here are some problems a city naturalist could solve:

1. How much of a bird's day is spent in looking for food?
2. How much time is spent in preening (grooming its feathers)?
3. How much time is spent in resting?
4. How often does the bird bathe?
5. How often does the bird drink?
6. How many different call notes are used by the male and by the female?
7. What animals cause fear reactions in the bird?

Collecting Feathers

Feathers are found in the greatest number from late July to early September. This is the season when birds are shedding their feathers (molting). The molting season follows the nesting season. Some species also molt before the nesting season.

A label can be attached to the shaft of the feathers you collect. Feathers can also be taped to index cards or cardboard and labeled. Different types of feathers from the same species can be put on one card. Many feathers you find will be hard to identify, but some will be easy. Color, size, and pattern help.

The barbs of feathers can be opened and closed easily. It is possible to see the tiny barbules with a magnifying glass or hand lens. The barbules work like hooks on a zipper. Watch a bird preen. It "zips" up its feathers by drawing them through its bill.

A Vacant Lot

COLLECTING INSECTS

A vacant lot where weeds, grasses, and perhaps a tree is growing is a good place to study insects. Ants, butterflies, grasshoppers, crickets, beetles, and caterpillars are common.

How To Get a Close Look at a Butterfly

All you need is a butterfly, sugar, patience, and some luck. Dissolve some sugar in your mouth. Put some of the sugary liquid on your finger. Move slowly toward a butterfly with your finger outstretched. If you are lucky, the butterfly will light on your finger. It may uncoil its drinking-straw mouthpart to suck up the sweet liquid. It is through this long mouthpart that the butterfly is able to reach down into flowers to suck up the nectar on which it feeds.

Tagging Monarch Butterflies

One of the butterflies you are likely to see is the black and orange Monarch. It lays its eggs on milkweed plants. When the caterpillars hatch, they feed on the plants.

The Monarch is one of the few insects that migrate. Hundreds of people are helping an entomologist (a scientist who studies insects) learn about the Monarch's migration by tagging the butterflies. A small patch of scales on the front right wing is rubbed off between the thumb and forefinger. A numbered sticker is pressed onto the bare spot.

Anyone finding a dead tagged Monarch is asked to pack it carefully and mail it to the address on the sticker together with a note telling where, when, and how the Monarch was found.

If you see many Monarch butterflies in your neighborhood, you may wish to help with this project. If you live in states bordering the Gulf of Mexico—Florida, Alabama, Mississippi, Louisiana, and Texas—your help is especially needed. Write to Professor Fred A. Urquhart, Scarborough College, West Hill, Ontario, Canada, for information.

How To Make an Insect Net

An insect net is a handy tool for catching insects. You can make your own net. You will need a nylon mesh shopping bag, 8 gauge wire, an old broom or mop handle 3 feet long, string, tape, a strip of cloth, a needle and nylon thread, and a drill.

1. Measure the width of the bag. Multiply by 2. Buy a length of wire 4 inches longer.
2. Drill a ¼″ hole in the handle 1 inch from the top. Drill a second hole 1 inch below the first. Drill the holes clear through the handle.
3. Bend the wire into a hoop.
4. Push one end of the wire through one hole and the other end through the other hole. Bend the ends of the wire so that 2 inches overlap along the sides of the handle.

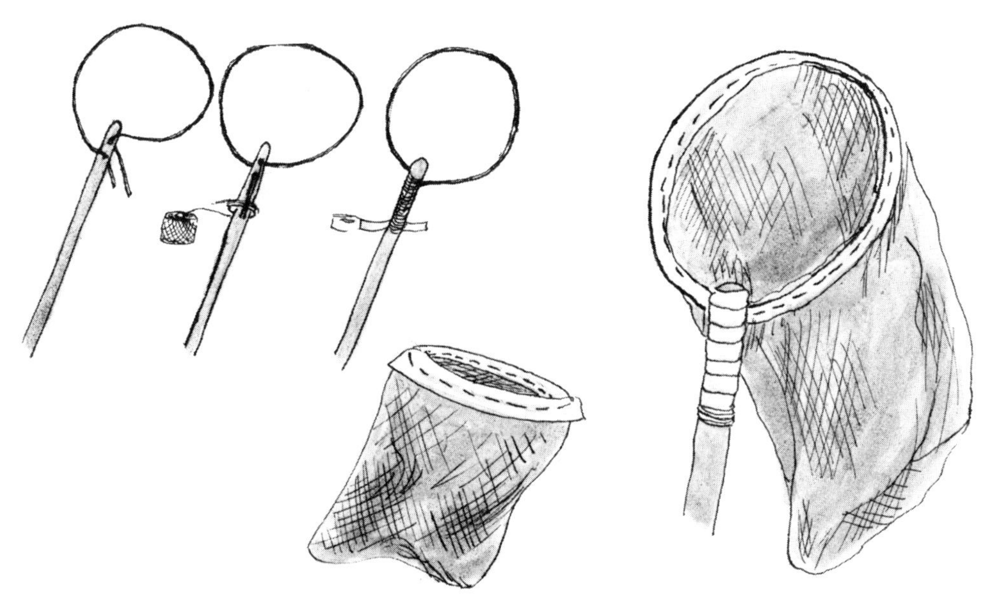

5. Use string to lash the wire ends tightly against the handle.

6. Wrap tape over the lashing to reinforce.

7. Cut off the handles of the shopping bag. Sew a 2-inch-wide strip of cloth around the top of the bag to reinforce it.

8. Fit the top of the bag over the wire hoop. Sew onto hoop with nylon thread.

Catching Insects With a Net

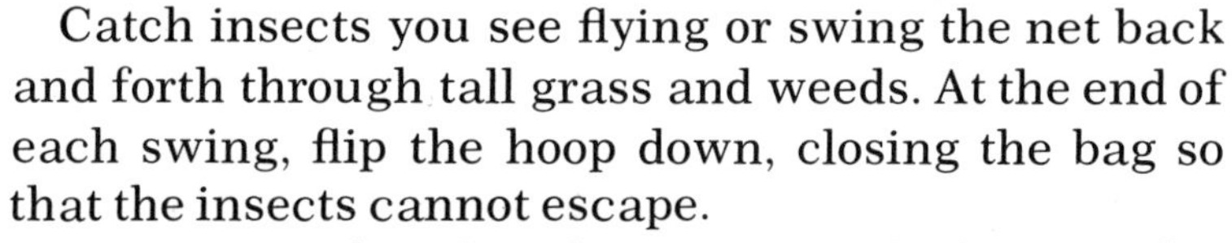

Catch insects you see flying or swing the net back and forth through tall grass and weeds. At the end of each swing, flip the hoop down, closing the bag so that the insects cannot escape.

Insects can be placed in a covered glass jar for viewing. A magnifying glass or a hand lens is needed to study small insects.

A magna vial is handy to use. This is a small plastic vial with screening in the bottom and a magnifying glass in the lid. Insects can be placed in the vial and observed through the magnifying glass in the lid as they move about. Magna vials are sold in museum shops or may be ordered from scientific supply houses. (See the list in back of book.)

Insects and spiders that sting or bite need to be stunned before they are taken from the net for viewing. This can be done by placing them, while still in the net, into a killing jar. Once they stop kicking, use tweezers or forceps to move them from the net into a viewing jar or magna vial.

Count the number of legs, the number of wings. What type of wings does the insect have? Note the color, size, shape, of the insect; the shape of the antennae; the type of mouthparts—chewing or other. Is there pollen on the insect's legs?

Once you have completed your observations, the insect can be released.

How To Make an Insect-Killing Jar

If you want to make a permanent insect collection, you will have to kill the insects you collect. To do this you will need to make two killing jars, one for moths and butterflies and one for other insects. The wing scales of moths and butterflies come off and stick to other insects placed in the same jar. This is why they should be killed separately.

To make the killing jars you will need two wide-mouthed jars with tightly fitting lids, plaster of Paris, water, a mixing can and stick, paper toweling, and nail polish remover. A small amount of plaster of Paris can be bought at an art supply or hardware store. If the lid has a cardboard lining, remove it before making the killing jar.

1. Pour one cup of water into a clean can.
2. Slowly add plaster of Paris to the water, stirring all the time, until it is as thick as heavy cream.
3. Pour mixture into clean jars to a depth of ¾ inch. Allow to dry 24 hours.
4. Place 3 or 4 strips of paper toweling in one jar

after the plaster is dry. The toweling will absorb moisture from the insect's body. The strips also help keep insects in the jar from clawing and biting one another. Do not put toweling in the butterfly and moth jar as too many wing scales from these insects would rub off on the paper.

5. Before starting out on a collecting trip, pour one tablespoon of nail polish remover on the plaster in each jar. Cover tightly. The fumes will knock out insects quickly, but keep the insects in the jar for 1 hour to be sure they are dead.

Mounting Insects

Insects in a collection are mounted on special insect pins. You will need to buy the pins. Number 3 size is best for most insects. The pins may be purchased from scientific supply houses. See the list on page 67.

Freshly killed insects are easier to mount than ones that have been dead some time and have dried out. Push an insect pin through the back of the insect's body, a little to the right of center. Leave about ½ inch of the pin sticking out to use as a handle. Add a label to the pin under the insect. Stick the pin in a large cork, styrofoam, or corrugated paper. Allow insect to dry. This may take two to three weeks.

How To Make a Spreading Board

Insects such as butterflies and damselflies must be dried on a spreading board so that the wings will be clearly visible when the insects are mounted on pins. Corrugated paper (12 x 12 inches) can be used as a simple spreading board.

1. Using forceps or tweezers, place the freshly killed insect on its back on the spreading board.
2. Pin insect to board through the thorax (chest).
3. Cut several strips of strong paper.
4. Spread wings apart with pins.
5. Using paper strips and pins, anchor wings in place.

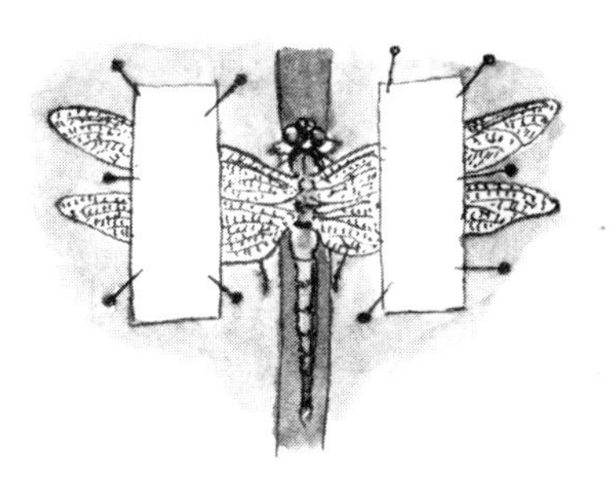

6. Carefully remove pin from thorax (leave pin in a dragonfly).

7. Allow three weeks for the insect to dry before removing it from the board.

8. Turn insect right side up and mount. (Carefully remove pin from a dragonfly and insert from the top.)

Caterpillars and other very soft-bodied insects for a permanent collection are killed and preserved in alcohol.

Keeping a Collection

Cigar boxes make excellent storage boxes for a collection of mounted insects. Line the bottom of the box with corrugated paper, styrofoam, or sheet cork. A teaspoon of moth flakes wrapped in a bit of old nylon stocking can be pinned in one corner of the box to keep other insects from eating your collection.

The directions given here are very simple ones. In the bibliography you will find a list of references which give detailed instructions for using the equipment and techniques of the entomologist.

Keeping Insects Alive

Live insects in cages are interesting to watch. Making daily observations can be fun. Once you have completed your observations, the insects can be released. Here are some suggestions for housing live insects.

Caterpillars

Caterpillars are colorful and easy to collect. When you find one, observe what it is eating. Most caterpillars eat leaves. Gather some of the same food. Wash it to remove any dirt and poisonous sprays that might be present. Place the food in a cage with the caterpillar.

To keep the food fresh, place it in a jar of water. Foil around the neck of the jar will prevent the caterpillar from falling into the water and drowning. Wrap the jar with tape so that the caterpillar can crawl back up the jar to the food if it falls. Fresh food should be provided daily.

Listen to the caterpillar eat. When does it eat? How much does it eat? How often does it molt (shed its skin)? Measure its growth. What changes do you see in the caterpillar?

The caterpillar will stop eating when it is about to pupate—change into an adult moth or butterfly. It should not be disturbed at this time. If your caterpillar is a moth caterpillar, it will spin a cocoon. If it is a butterfly caterpillar, it will form a chrysalid.

Cocoons and Chrysalids

The cocoon or chrysalid can be placed in a jar as shown. Keep a blotter or sponge wet to provide moisture. Tape a strip of paper toweling to the inside of the jar so the adult moth or butterfly will have something to cling to when it emerges. Cover the jar with a piece of old nylon stocking held in place with a rubber band.

Watching the adult emerge is very exciting. Most butterflies and moths emerge in the spring, usually in the morning. As soon as the adult is able to fly, release it. If it is a moth, release it in the evening as moths are active at night.

Even if you do not raise a caterpillar, you may find cocoons and chrysalids to collect. The spraying of trees and shrubs has caused many of the larger butterflies and moths to disappear from our cities. The caterpillars are killed when they eat leaves sprayed with poison. One cocoon that may still be found in large numbers in cities is the bagworm cocoon.

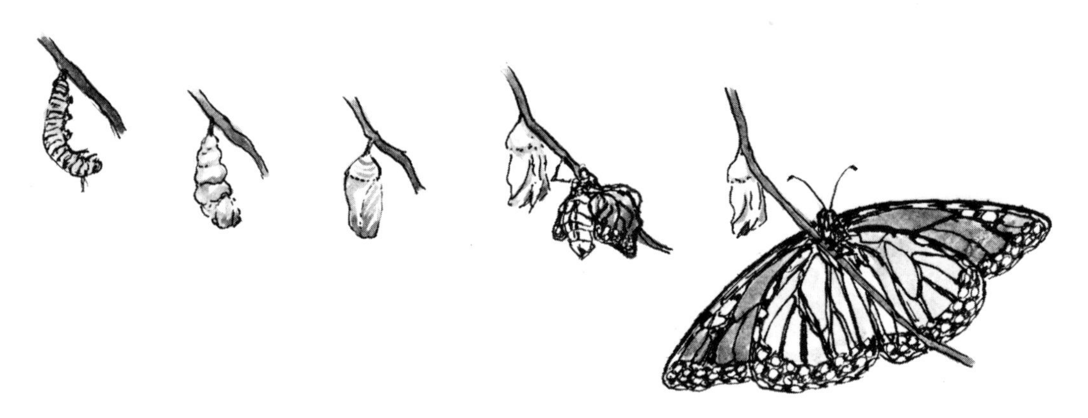

Bagworms

In spring the newly hatched bagworm caterpillar weaves a silk bag about itself. It adds bits of leaves from the trees and shrubs on which it feeds to the silk bag. The bag is carried about by the caterpillar all summer long and is made larger as the caterpillar grows.

In the fall the caterpillar attaches the bag to a twig, lines the inside of the bag with silk, and pupates. The adult emerges about three weeks later.

The adult female does not have wings and cannot fly. The adult male flies to the female and dies after mating with her. The female lays her eggs inside her bag, then she too dies. The eggs remain in the bag all winter and hatch the following spring.

Collect a few young bagworm caterpillars. Place them in a large jar. Use a piece of old nylon stocking for a lid. Feed the caterpillars lettuce. Watch to see if bits of lettuce are added to their bags.

Crickets

Another interesting insect to watch is the cricket. Crickets can be found under stones and boards. An empty fish tank, enamel pan, or a plastic box with

two inches of sand or soil and a few leaves makes a good cage. The soil or sand should always be slightly damp. Crickets are good jumpers so a fine mesh wire cover is needed.

Crickets can be fed lettuce, bits of dog biscuit, bread, hard boiled egg, thin slices of apple, cucumber, or carrot. Remove uneaten food every other day. Sprinkle a few drops of water on the side of the cage each day for the crickets to drink.

When you collect crickets, try to collect both males and females. The female cricket has a long ovipositor for laying eggs underground.

By watching crickets you may learn whether both sexes chirp. How is the sound made? What amount of light makes them active? At what temperature are they most active?

Ants—How To Make an Ant House

Housing for ants can be quite elaborate as ants live underground and feed above ground. But a simple cage can be made from a large glass jar with a lid and some black paper. A gallon jar is a good size to use.

You will have to make a few holes in the lid with a nail and hammer. Punch the holes from the inside of the lid out, then hammer the jagged edges flat so that you won't get cut when you screw the lid on and off. Line the inside of the lid with fine mesh wire so that the ants will not escape through the holes.

Collecting Ants

To collect ants you will need a trowel, a piece of white cloth, a spoon, and two empty cans with lids. The two-pound size is best.

1. Look in a vacant lot for a nest of ants. Pick an open area where the digging will not be too difficult.
2. Starting a foot away from the ant nest, use the trowel to scoop up soil and ants. Dump a scoopful at a time on the white cloth to see the ants better.
3. Gently spoon the ants into one of the collecting cans.
4. You will also want to collect the queen. She will be larger than the other ants in the colony. You may have to dig deep, exposing the tunnels and underground chambers to find her. If you do, place her in the collecting can with the other ants. If you do not, the colony will survive anyway for a while in your ant house.
5. Before filling in the hole you dug, fill the other can with soil.

At home, fill the gallon jar ¾ full of soil. Tape black paper around the outside of the jar. Dampen a piece of sponge with water. Place it on the soil. Dissolve some sugar in a teaspoon of water. Put the sugar syrup in a bottle cap on the soil for the ants to eat. Gently transfer the ants from the collecting can to the jar. Put on the lid.

Keep the black paper around the jar except for brief periods when you want to watch the ants "underground." Do not remove the black paper for very long, or you will disturb the colony.

The ants will make tunnels and chambers in the soil. You may be able to watch them care for their eggs, the cocoons in which they pupate, and other activities of a busy ant colony.

Experiment with feeding the ants bits of raw meat, bread crumbs, jam, candy, and other foods. The ants should be fed twice a week. If the ants try to escape when the lid is removed, use a small paintbrush to pick them up and return them to the jar. Always keep the sponge moist.

Trapping Ants

If you cannot find a suitable ant nest to dig up, you can try to trap a colony of house ants.

1. Lay a glass jar on its side. Fit it with a cardboard roof to shut out light. Ants like dark places.
2. Place a wet sponge in the jar. Ants like damp places.
3. Make a paper ramp so the ants can walk into the jar.
4. Bait the trap with a few drops of sugar syrup.
5. Set the trap where ants have been seen. During the night the whole colony may move in with their queen and eggs and young.

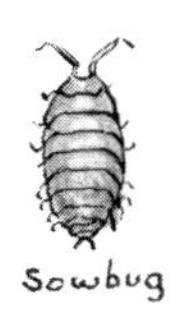

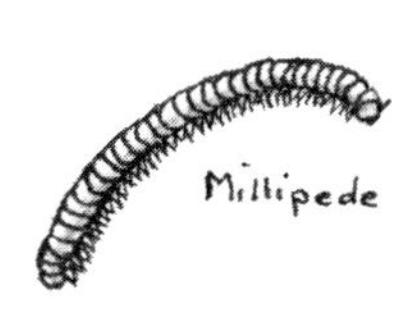

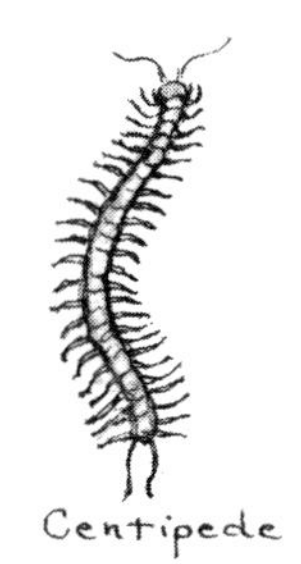

Other Animals in a Vacant Lot

Many small animals other than insects can be found in a lot. You may find earthworms, sowbugs, spiders, centipedes, millipedes, and snails. Lift up the litter, rocks, and wood. The animals may be hiding underneath.

Land Snails

All of these animals are easy to keep, but land snails are perhaps the most interesting. Place them in a large glass jar. You will need to punch a few holes in the lid with a hammer and nail. Be sure to punch the holes from the inside of the lid out, and then to hammer down the jagged edges of the holes.

Feed the snails lettuce or leaves from the plant on which you found them feeding. Watch how they eat,

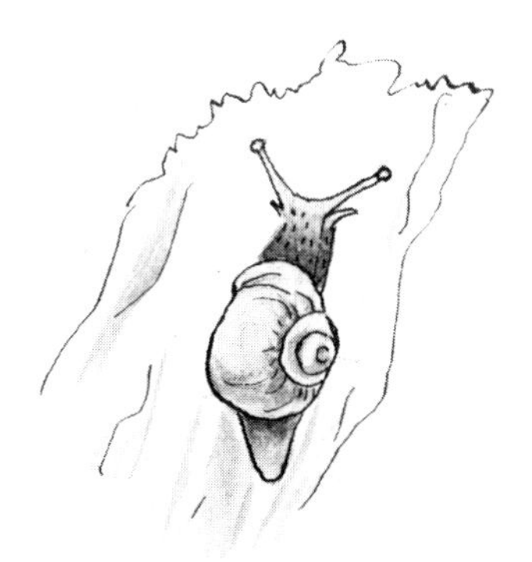

move about, how they use their feelers and their eye-stalks.

Keep a moist paper towel in the bottom of the cage. If the cage becomes dry, a thin layer of mucus will harden across the shell opening and the snails will remain inactive.

Snails require calcium to keep their shells hard and to allow their shells to grow. If you keep the snails more than a few days, you should put an empty snail or sea shell in the cage for the snails to feed on. A cuttlebone, sold in pet stores, can be used in place of a shell to supply the needed calcium.

Collecting Spider Webs

Flat, wheel-shaped webs that you find out-of-doors can be collected. A jar with an atomizer top, paint, newspaper, plain paper, and scissors are needed. Dark-colored paper and light-colored paint are best. It may be necessary to dilute the paint so that it will pass through the atomizer.

If the spider is on its web, frighten it off with a twig. Do not use your fingers as spiders bite. Some spider bites can be serious.

1. Protect everything around the web with newspaper before starting to spray.
2. Spray both sides of the web with paint. Use lots of paint. Hold the paint jar at an angle so as not to break the web with the full force of the spray.
3. Ease the plain paper close to the underside or

back of the web. Carefully touch the paper to the whole web at once.

4. Holding the paper in one hand, cut the guy lines which attach the web to the tree or bush.

5. Set paper aside to dry. Label.

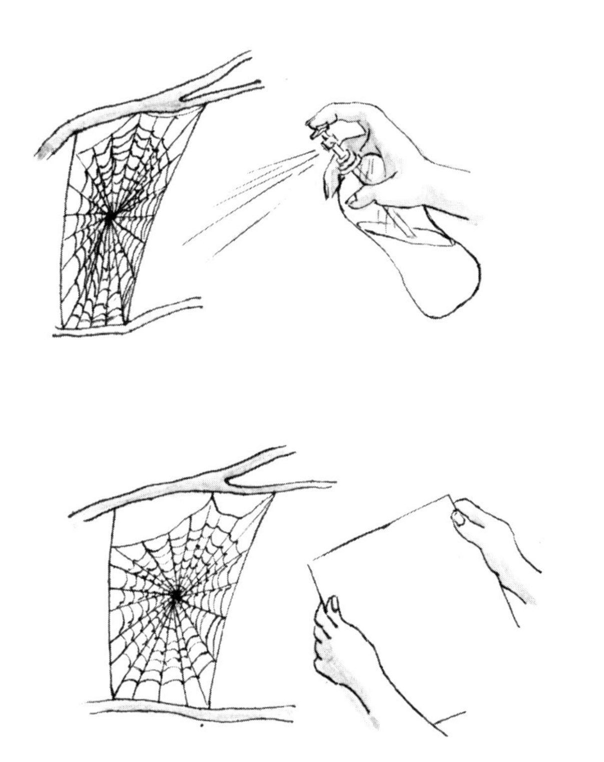

A Pond in the Park

COLLECTING POND LIFE

The park pond is probably the only place left in the city where you can study pond life. When you think of a pond and its animal life, you probably think first of fish, frogs, and turtles. Unfortunately, many of these larger animals have disappeared from city ponds. The ponds are no longer "wild" enough to support them or the pond may be polluted.

It is surprising though how many animals still inhabit city ponds. You might try exploring for these less familiar animals. If they are collected in limited numbers, observed, and then released in the pond where they were found, no one should object to your collecting them.

Equipment

Pond exploring tools can be simple. Use a dip net or a kitchen strainer to sweep through the water. For a longer reach, lash an old broom or mop handle to the strainer. To make a sieve for scooping up the bottom of the pond, tack fine mesh wire to two strips of wood.

A few sweeps of the net or strainer, some scoops of bottom muck, a careful examination of the stems

and leaves of water plants, should yield a number of interesting animals.

To examine your catch, put them in a pie tin with a half-inch of pond water. As a few aquatic insects bite, use forceps or tweezers in handling them.

It will be impossible for you to identify all the animals you catch and to learn their habits in one afternoon, so don't try. Keep only a few to observe further. Return the rest to the pond to collect another day.

Use a pail to hold your catch. Half fill the pail with pond water. Add two or three pond plants for the animals to hide among. In the pond the plants shelter the animals from predators (enemies). Keep the pail out of the sun until you start home. If the water gets too warm, the animals will die.

Here are several different pond animals to look for and suggestions for housing them.

Diving Beetles

A diving beetle can be kept in an empty fish tank or wide-mouthed gallon jar.

Put a layer of clean sand in the bottom. Add pond water and one or two pond plants. After the water clears, add the beetle. It bites, so use forceps. Bend fine mesh wire to make a cover. Feed the beetle insects or bits of raw meat. Remove any uneaten food so as not to foul the water.

While observing the beetle, see if you can learn the answers to the following questions. What is the resting position of the beetle? Why? How long can it remain underwater? Why does it have flattened hind legs? How does it capture its food? What role does it play in the pond?

Return the beetle to the pond in two or three days when you have completed your observations.

Dragonfly Nymphs

An enamel or plastic pan, half-filled with pond water, can be used for a dragonfly nymph (a young dragonfly). Add a few water plants to which the nymph can cling. Feed it bits of raw hamburger. Use tweezers, a broomstraw, or toothpick to wiggle the meat in front of the nymph. Remove any uneaten food.

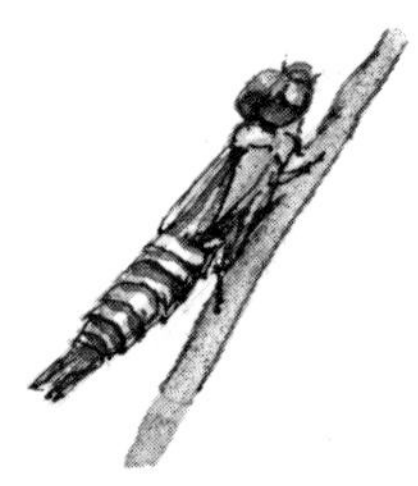

If the nymph tries to climb out of the water it may be ready to change into an adult dragonfly. Using a flower holder or florist's clay, anchor a small stick to the bottom of the pan for the nymph to climb when it is ready to shed its skin. You will be in for a rare treat if you see the adult emerge.

When the adult dragonfly emerges, let it fly away to catch its own food (mosquitoes).

Pond Snails

Pond snails are easy to keep. Look for them on the stems and undersides of leaves of water plants and under stones on the pond bottom.

Do not mistake land snails (which sometimes come to the pond edge) for pond snails. To tell them apart, count the tentacles on their heads. Land snails have four tentacles, pond snails have two.

Use a fish tank or wide-mouthed gallon jar to hold the snails. Put in a layer of clean sand and nearly fill the tank or jar with pond water. Add a few pond plants. Let the water clear before adding the snails.

The snails will feed on the plant stems and on algae that may grow in the aquarium. To be sure they have enough food, add a piece of lettuce once a week.

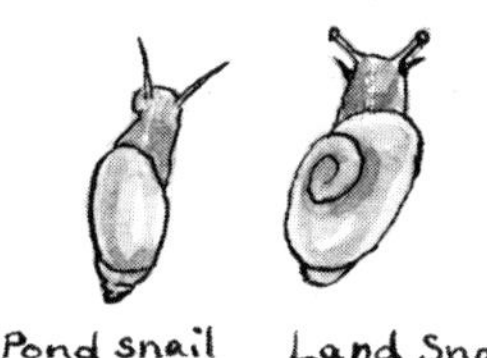

A few kinds of snails bear living young, but most lay eggs. In a pond the eggs are usually laid on plants or stones. In a glass tank, snails often lay their eggs on the sides of the tank. The eggs are surrounded by clear jelly. You can watch the young snails grow. A hand lens or magnifying glass will help you to see the changes that take place.

Crayfish

Crayfish are active at night. During the day they are apt to remain hidden and have to be prodded out of their shelters with a stick. If you are quick, you can catch hold of a crayfish by the middle of its back, but it is safer to use a dip net or strainer. Crayfish pinch.

Keep only one crayfish at a time. Two together may fight. Put the crayfish in a tank of pond water and watch it walk and swim. How many legs does it have? How do the legs differ? How are they used? How does it use its antennae?

To make a temporary home for a crayfish, add three inches of clean coarse gravel to the bottom of a small tank. The crayfish will make a den in the gravel. Half fill the tank with pond water. Place a

stone in the middle so the crayfish can crawl out of the water. Cover the tank with wire mesh to prevent the crayfish from escaping. Put a few pond plants in the aquarium, but be prepared for the crayfish to eat them.

Change the water whenever it is dirty. Use water from the pond, if possible. Let it clear before adding it to the tank. If you must use tap water, allow it to stand two days in a non-metal container before adding it to the tank. This is to allow any chlorine that has been added to the water to purify it to escape.

Feed the crayfish small pieces of fish and meat two or three times a week. Wiggle the food in front of the crayfish using forceps or tweezers. Remove any uneaten food.

Because the crayfish's shell does not stretch, the crayfish must shed its shell in order to grow. The shell breaks open across the back. The soft-bodied crayfish pulls itself out of its old shell and remains hidden until the new shell hardens.

Tadpoles

Place one to four tadpoles in an enamel pan or plastic basin with three inches of pond water. Feed the tadpoles a piece of raw spinach or a piece of lettuce. Do not feed again until the food has been eaten. Change the water when it looks dirty, using water from the pond or tap water that has stood for two days in a non-metal container.

Gradually the tadpole's tail will be absorbed and the tadpole will grow four legs. When it does, place a piece of wood or a rock in the pan so the tadpole will be able to get out of the water. At this stage the tadpole can jump, so cover the pan with wire mesh.

There will be changes going on inside the tadpole that you will not be able to see, but may be able to think out for yourself. For example, when the tadpole becomes an adult, it is no longer a plant eater, but is a meat eater. Changes must take place in its digestive system for this to happen.

When the tadpole has changed into a frog or toad, return it to the pond (to the edge of the pond if it is a toad). Toads can be told from frogs by the two large paratoid glands behind the eyes.

Some tadpoles transform, or change, within a few weeks. Others take longer. Green frog and bullfrog tadpoles take two and sometimes three summers to transform into adults. You may not wish to keep the tadpoles that long, so return them to the pond.

When exploring ponds you may find egg masses laid by frogs or toads. These should *not* be collected as many eggs would spoil and never hatch. In time, this would deplete the numbers.

How to Cast an Animal Track

Around the edge of the pond where the ground is soft, you may find the tracks of animals—dogs, pigeons, cats, sparrows, worms, toads. To make a permanent record of the tracks you will need plaster of Paris, water, mixing can and stick, a strip of cardboard one and a half inches wide, and paper clips.

1. Find a track that is clear and sharp.
2. Clear the ground around the track, carefully removing any sticks, leaves, stones.
3. Surround the track with the cardboard strip. Fasten the ends of the strip together with paper clips. Push the cardboard "collar" down lightly into the ground.

4. Into the mixing can put as much water as will be needed to fill the cardboard collar.

5. Add plaster of Paris to the water, stirring all the time until the mixture is as thick as heavy cream.

6. Pour mixture slowly over track.

7. If you want to hang the cast when it is finished, slip an opened paper clip into the cast near the top before the plaster hardens. Let the loop extend above the plaster to serve as a hanger.

8. After the plaster hardens (30 minutes) lift it from the ground. Allow the cast to dry overnight.

9. Remove the collar. Carefully brush off any remaining soil. The cast is complete.

The cast will be the exact reverse of the track. Where the track went in, the cast will stick out. It is called a negative cast.

To make a positive cast, lightly grease the negative cast with petroleum jelly or cold cream. Put a collar around the cast and fill the collar with freshly mixed plaster. When the plaster hardens, remove the collar and separate the two casts, negative and positive. The positive cast looks just like the original track.

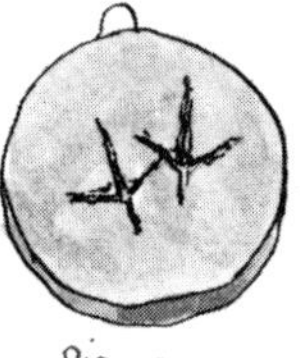
Pigeon

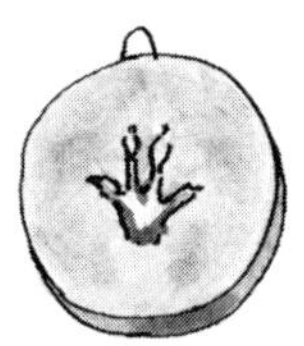
Squirrel (front foot)

Cat

A Selected Bibliography

In addition to the books listed here, the naturalist living in a city has many other sources of help close by—natural history museums, natural science centers, botanical gardens and arboretums, zoos, aquariums, libraries, natural history clubs and societies.

Insects

The Adventure Book of Insects by Alice Gray, Capitol Publishing Company Inc., New York, 1956.

Caterpillars by Dorothy Sterling, Doubleday & Company, Garden City, New York, 1961.

Collecting Cocoons by Lois J. Hussey and Catherine Pessino, Thomas Y. Crowell Company, New York, 1953.

Crickets by Olive L. Earle, William Morrow and Company, New York, 1956.

The Insect Guide by Ralph Swain, Doubleday & Company, Garden City, New York, 1949.

Insects in Their World by Su Zan Swain, Garden City Books, Garden City, New York, 1955.

Pond Life

Pets from the Pond by Margaret W. Buck, Abingdon Press, New York, 1958.

Birds

The Bird Watcher's Guide by Henry Hill Collins, Jr., Golden Press, New York, 1961.

Birds of North America by Chandler S. Robbins, Bertel Bruun, and Herbert S. Zim, Golden Press, New York, 1966.

Plants

Learn the Trees from Leaf Prints by David S. Marx, The Botanic Publishing Company, Cincinnati, Ohio, 1940.

The Tree Identification Book by George W. D. Symonds, M. Barrow and Company, New York, 1958.

Weeds by Dorothy C. Hogner, Thomas Y. Crowell Company, New York, 1968.

Wild Green Things in the City by Anne O. Dowden, Thomas Y. Crowell Company, New York, 1972.

Geology

Collecting Small Fossils by Lois J. Hussey and Catherine Pessino, Thomas Y. Crowell Company, New York, 1970.

The Story of Rocks and Minerals by David M. Seaman, Harvey House, Irvington-on-Hudson, New York, 1956.

General

A Crack in the Pavement by Ruth Howell, Atheneum, New York, 1970.

Exploring As You Walk in the City by Phyllis S. Busch, J. B. Lippincott Company, New York, 1972.
Science in a Vacant Lot by Seymour Simon, The Viking Press, New York, 1970.
Small Worlds by Helen Ross Russell, Little, Brown and Company, Boston, 1972.
Golden Nature Guides, Golden Press, New York. Titles in this series include *Birds, Butterflies and Moths, Flowers, Fossils, Insects, Pond Life, Reptiles and Amphibians, Rocks and Minerals, Trees, Weather.* Golden Regional Guides and Golden Science Guides are other series published by Golden Press.
Peterson Field Guides, Houghton Mifflin Company, Boston. Titles in this series include *Animal Tracks, Birds, Butterflies, Rocks and Minerals, Trees and Shrubs.*

A large number of excellent recordings of animal sounds are available including *A Field Guide to Bird Songs*, Houghton Mifflin Company, Boston; *The Songs of Insects*, and *Voices of the Night* (frogs and toads), Laboratory of Ornithology, Cornell University, Ithaca, New York.

SCIENTIFIC SUPPLY HOUSES

Carolina Biological Supply Company (at either address)
Burlington, North Carolina 27215
Gladstone, Oregon 97027

General Biological Supply House
8200 South Hoyne Avenue, Chicago, Illinois 60620

Ward's Natural Science Establishment (at either address)
P. O. Box 1712
Rochester, New York 14603
P. O. Box 1749
Monterey, California 93940

Damon Educational Division (to order a magna vial)
80 Wilson Way
Westwood, Massachusetts 02090

INDEX

ABOUT THE AUTHORS

Lois J. Hussey was first attracted to natural history during her childhood in Oyster Bay, Long Island. Her interest grew and she went on to graduate from Adelphi University with a degree in biology. She was able to develop her interests when she took a position with the Department of Education of the American Museum of Natural History, where, for more than twenty years, she engaged in teaching and administrative work. In addition, she found time to write. She is co-author, with Catherine Pessino, of *Collecting Cocoons* and *Collecting Small Fossils*. In private life Mrs. Douglas E. Heilbrun, the author now makes her home in Flushing, New York.

Catherine Pessino is a native New Yorker. She was graduated from Hunter College with a degree in biology, and then joined the Department of Education of the American Museum of Natural History. At present, Ms. Pessino is Supervisor of the Museum's Alexander M. White Natural Science Center, an exhibit/teaching area for young people on the ecology of New York City. In 1971, Ms. Pessino received the Elsie M. B. Naumburg Award, bestowed by the Natural Science for Youth Foundation for outstanding work with children in the field of natural science. She has participated in several exciting research projects for the Museum. She spent two summers in Alaska with the tundra Eskimos and more recently she helped to establish an extensive research program studying two species of tern on Great Gull Island in Long Island Sound.

ABOUT THE ILLUSTRATOR

Barbara Neill has been in the museum and nature-center field for more than twenty years. Now a colleague of Catherine Pessino in the Alexander M. White Natural Science Center at the American Museum of Natural History, she has made a full circle, for she began her career as a volunteer in that museum. In the years between she traveled to Charlotte, North Carolina, to become an assistant in the children's museum, then to the Natural History Museum of Santa Barbara, California, then back east to Connecticut where she was first director of the New London County Children's Museum, and then director of the Lutz Junior Museum in Manchester, Connecticut. In addition to her work at the Museum of Natural History, she spends as much time as possible drawing and painting animals at her home in Tenafly, New Jersey.